Application of ISO/IEC 17025 Technical Requirements in Industrial Laboratories

METHOD VALIDATION

by M.L. Jane Weitzel and Wesley M. Johnson

Copyright © 2013 by M.L. Jane Weitzel and Wesley M. Johnson
First Edition –February 2013

ISBN
978-1-4602-1029-1 (Hardcover)
978-1-4602-1027-7 (Paperback)
978-1-4602-1028-4 (eBook)

All rights reserved.

No part of this publication may be reproduced in any form, or by any means, electronic or mechanical, including photocopying, recording, or any information browsing, storage, or retrieval system, without permission in writing from the publisher.

Produced by:

FriesenPress
Suite 300 – 852 Fort Street
Victoria, BC, Canada V8W 1H8

www.friesenpress.com

Distributed to the trade by The Ingram Book Company

Table of contents

INTRODUCTION 1

Method validation – what is it? 1

Method validation – why is it necessary? 1

Method validation – when should it be done? 2

Relation of method validation to method verification 2

USING AN APPROPRIATE METHOD 3

What is method development? 3
    Measurand 4

What is method selection? 5
    Fitness for purpose 5
        Data concepts that impact fitness for purpose 6
            Accuracy and Uncertainty Conundrum (AKA Accuracy and Precision Trade-off) 6
            Reportable Value 7
            Sampling 11

Decision rule approach and target uncertainty 11
    Measurement uncertainty 11
    Decision rules 12
    Determining the target uncertainty 13
        How to determine a target measurement uncertainty 13
        Impact of Bias and Uncertainty Conundrum on Target Uncertainty 15
    Decision rules approach: 16
        Using the N:1 decision rule 16
        Simple decision rule for an upper specification limit that includes the uncertainty 17

Simple decision rule for an upper specification limit when the
uncertainty is not given ... 17
    What does probability mean? ... 18
    The process of using probability to develop a decision rule for an upper specification limit when the uncertainty is not given ... 20
    Setting target uncertainty when specification states the guard band ... 22
Uncertainty and cost approach ... 24

METHOD VALIDATION ... 24

Method validation pre-requirements ... 25
Instrument qualification ... 25
    1. Sample form, whether it is liquid or sample ... 26
    2. Sample size ... 26
    3. Range of the instrument ... 26
    4. Instrument stability ... 27
    5. Instrument interferences ... 27
    6. Precision ... 27
    7. Instrument response to concentration ... 27
    8. Cost of operation ... 27
Computer and software qualification ... 27

Management of method validation ... 28

Overview of validation activities ... 32
Extent of validation ... 33
    Validation categories ... 33
        Category 1: Identification ... 35
        Category 2: Analyte at low concentration, quantitative ... 35
        Category 3: Analyte at low concentration, limit test ... 35
        Category 4: Analyte at high concentration, quantitative ... 36
        Category 5: Analyte at high concentration, limit test ... 36
        Category 6: Qualitative ... 36
Validation tools ... 36
    Reference materials ... 36
    Blanks ... 38
    Interference checks ... 38

Replicates, especially duplicates	39
Duplicates	39
Design of experiments and statistics	40
MS EXCEL as a validation tool	42

Validation activities — 42

Selectivity/specificity	42
Purpose	42
Analytical activity	42
Acceptance criteria	43
Discussion	43
Limits of detection and quantitation	44
Purpose	44
Limit of detection	44
Limit of quantitation	44
Analytical activity	45
Limit of detection - quantitative	45
Limit of detection – qualitative	45
Limit of quantitation	46
Acceptance criteria	46
Discussion	46
Measuring interval (range)	47
Purpose	47
Analytical activity	47
Use of reference samples or spikes	48
Instrument response	48
Linear interval	48
Acceptance criteria	49
Discussion	49
Accuracy, trueness and bias	50
Purpose	50
Analytical activity	52
Acceptance criteria	54
Discussion	54
Precision (repeatability, intermediate precision and reproducibility)	56
Purpose	56
Repeatability	56
Intermediate precision	57

Reproducibility	57
Analytical activity	57
Repeatability	57
Intermediate precision	57
Reproducibility	58
Use of ANOVA to estimate the various precisions	58
Duplicates	60
Acceptance criteria	60
Discussion	60
Ruggedness (or robustness)	60
Purpose	60
Analytical activity	61
Ruggedness test	61
Discussion	63
Measurement uncertainty	64
Purpose	64
Analytical activity	64
Bias	65
Precision	66
Discussion	66

CONCLUSION 68

REFERENCES 69

Method Validation

Introduction

Method validation – what is it?

Method validation is the process of performing a series of experiments designed to prove that the measurement procedure to be used for an analytical task meets the performance criteria agreed to between the laboratory and the laboratory's client. "In VIM 3, validation is the verification or check that the 'given item', e.g. a measurement process … is fit for the intended purpose."[i] This would involve a series of experiments that would assess performance criteria such as trueness, precision, measuring interval (or measuring range), limit of detection and of quantification, selectivity, bias, confirmation of identity, recovery, sensitivity, ruggedness and measurement uncertainty.

Method validation – why is it necessary?

Modern society is very dependent on analytical work in the daily conduct of business, in purchasing decisions, in health issues, the regulation of food and pharmaceuticals, mineral exploration and extraction, forensic investigations and many more too numerous to list. In each case decisions of one sort or another have to be made and the reliability of the decision is dependent on the quality of the analytical result. The methods used in these various applications will be dependent on the measurand, the sample matrix and many other factors. In all cases however, the method must be shown to be capable of generating data that is fit for the purpose to which it will be put. In order to assure the client that the data is fit for purpose the method must be validated.

The various decisions to be made could involve large sums of money (billions of dollars to develop a mine for example), someone's health, a criminal conviction or some other critical determination. Whatever the situation, the reliability of the analytical result is an important factor in reaching a final conclusion that minimizes the risk of being incorrect.

Method validation– when should it be done?

There are several times in a method's life when it should be validated; a method validation is not forever.

When a new method is developed and before it is used to generate results for clients the ISO/IEC Standard 17025[ii] requires that it be fully validated for the appropriate method criteria to assure that it is fit for purpose.

When an established method is revised or its application is extended past the parameters of a previous validation the method should be revalidated for the method parameters affected by the change.

When the routine monitoring of QC data indicates a change of method results with time and the root cause investigation identifies the method as a root cause, the laboratory should revisit the validation of the method.

Relation of method validation to method verification

Not all industries treat validation and verification as synonymous terms. Some use them differently with validation referring to proving a method is fit for purpose. Verification, on the other hand, is confined to proving that an already validated method can be used successfully in a laboratory. This is the distinction that will be made in this book.

Using an appropriate method

There is a saying "A carpenter is only as good as his or her tools." The same holds for analytical chemists. A chemist is only as good as the method used. ISO 17025-2005 in section 5.4.2 states in the first line that "The laboratory shall use test and /or calibration methods, including methods for sampling, which meet the needs of the customer and which are appropriate for the tests and/or calibration it undertakes." This means the laboratory must take care when either developing or selecting a method.

What is method development?

Method development is a complex topic that is addressed in many sources. Method development is not in the scope of this book. This section will address how to define the purpose of a method so that the method's performance parameters are clear.

Method development begins with the description of the use for the method. Then the technique is chosen and the analytical procedure developed. The same acceptance criteria used for validation are the goals for the method development. These acceptance criteria are discussed in detail in the method validation part of this book.

The capabilities of a test method should be established before a method is developed. Then the laboratory knows how "good" the method needs to be and does not waste money on developing a method that has an uncertainty much smaller than is needed or develops a method that is not good enough.

Laboratories supporting different industries may take different approaches to method validation. The documentation is more rigorous and formal in some industries than in others. The method validation activities may occur concurrently with the method development. However, the essential elements of method validation are the same, regardless of the industry.

Method development is often an iterative process in that the performance criteria for the test method change as the method is developed. For example, during development of a new pharmaceutical product, its formulation may be altered and the test method being developed must be modified to match the new requirements. In the mining industry the samples may come from new ore bodies when they are brought into production and the test method, especially the sample preparation step, may need to change to accommodate differences such as larger particle size or different concentration ranges. Whenever the sample type or the specifications vary, the purpose of the method changes and must be revisited to ensure the method that is being developed or modified will be fit for purpose.

The laboratory has to validate a method that has been developed before using the method to test client samples.

Measurand

The measurand will assist in defining the purpose of the method because the measurand is a specific description of what is being measured. When the measurand is defined the following parameters and conditions should be considered:

- The analyte, which is the component being determined or measured. For example, the measurand may be the concentration of lead in drinking water in which the analyte is lead.
- The value being measured, such as amount of substance or mass or concentration.
- The sample matrix.
- A definition of whether the result will refer to the laboratory sample or to the parent body from which the sample was taken, for example, a shipload of grain.

- The method being used if the result obtained depends upon which procedure is used. This is also known as an empirical method or operationally defined method.
- The identification of all important parameters or conditions that could impact the result. For example, the determination of water by loss on drying should specify the drying temperature if different results are obtained at different temperatures.

What is method selection?

The laboratory can use existing methods as long as they meet the needs of the customer. It is preferable to use methods that have been published by reliable institutions such as AOAC, United States Pharmacopeia (USP), European Pharmacopeia (EP), American Society for Testing and Materials (ASTM), etc. These are often known as "standard methods". They have been validated and demonstrated to be suitable for use by the institutions. The most recent version of these methods must be used, unless it is not appropriate. In many instances an appropriate standard method is not available, in which case the laboratory has to develop its own method.

The laboratory has to confirm it can perform the standard method adequately before using the method to test client samples. In some industries this is called method verification.

Fitness for purpose

The purpose of the method must be stated in such a way that the required analytical parameter, the bias, uncertainty, precision, specificity, etc. can be defined. The target values for these parameters form part of the "fitness for purpose" statement for the method, otherwise known as the method validation acceptance criteria.

The requirements for a method are often stated descriptively using phrases such as those that follow:

- The test result must be sufficiently accurate...
- Any bias must not be such that an incorrect decision will be made.
- The variability of the test result must be small enough so that the decision made...
- The result is based on the data and is not based on random variability.
- The test method must be sufficiently sensitive so that the decision an analyte is present or absent is correct and not solely based on the method's capability.

These are qualitative words that describe the concept, but are not the information needed to design a method validation or method verification. The laboratory needs a quantitative description of "fit", "adequate", "small enough", and "sufficiently". Then the quantitative acceptance criteria can be assigned for the analytical parameters such as bias, precision, or ruggedness.

The challenge comes in defining "fit" and "adequate". Measurement uncertainty (MU) is a key characteristic to clearly, simply, and unequivocally meet that challenge. The approach this book takes is that the target measurement uncertainty is known and is included in the fitness for purpose statement. The term target measurement uncertainty means the largest measurement uncertainty a method can have and still be fit for purpose.

Data concepts that impact fitness for purpose

Accuracy and Uncertainty Conundrum (AKA Accuracy and Precision Trade-off)

The characteristics of accuracy and precision have to be considered together when assessing fitness for purpose. The larger the bias in the results obtained with a method, the smaller the precision must be so that the range of values obtained for a sample will not overlap the limit just due to random variation. Regulatory bodies and standards groups have or are starting to address this issue in their analytical requirements. The United States Pharmacopeia (USP) has proposed using Operational Characteristic Curves to ensure that the combination of bias and precision is adequate.

Reportable Value
It is important to understand the concept of reportable value. Methods often include some replication. At least a part of the method is executed more than once, in duplicate, triplicate, or in a higher number of replications. The replication can occur at any step in the method. The replication must be well defined and consistent; the description of the reportable value identifies the number of replicates and at which step the replication occurs.

The reportable value is derived from one full execution of the method. The full execution of the method starts with a test portion of a laboratory sample, continues with sample preparation, instrument readings, etc., and includes the averaging of the individual values from any replication. This average is the reportable value. It is this reportable value that is compared to the specification. For example, if the method calls for two sample preparations and these two preparations are each presented to the measuring instrument twice, then the four individual values are averaged; this average value is the reportable value.

The reportable value is based on a statistical concept, the central limit theorem, which states that the distribution of sample averages tends to be a normal distribution. This normal distribution of averages holds true regardless of the distribution of the population of individual values. The standard deviation of the sample averages is called the standard error of the mean (SEM) and is calculated from the standard deviation, s, of the individual values based on the number of values, n, used to calculate the average:

$$SEM = \frac{s}{\sqrt{n}} \qquad \text{Equation 1}$$

The acceptance criterion for the standard deviation of the reportable value, the standard error of the mean, is based on the replication specified in the method; if the replication changes, the acceptance criterion will no longer be valid.

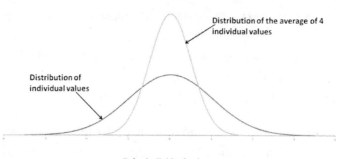

Figure 1 The distribution of the average of 4 values is described by the standard error of the mean which is ½ the standard deviation of the individuals. The standard error of the mean is $s/\sqrt{n}$, where s is the standard deviation of the individuals and n is the number of replicates used to calculate the average.

In method validation, the relation of the standard error of the mean to the standard deviation can be leveraged to reduce the amount of testing or number of experiments. The precision can be determined using individual values and then the precision of the reportable value can be calculated as the standard error of the mean using the standard deviation for the individual result and n, the number of replicates in the reportable value. For example, if the reportable value is obtained from the average of 4 individual values, the repeatability, which is the standard deviation of 7 repeats, can be determined from 7 replicates; the lab does not have to perform 7 x 4 = 28 replicates. This is useful in circumstances such as having limited quantities of reference material or test sample, or when the method will be implemented using different reportable values such as when better precision is needed to monitor a production process than is needed to release the final lot.

During routine use of the method, the replication for deriving the reportable value must be used so that the method performs as defined in the fitness for purpose statement. In the example in Figure 1 above, 4 replicates would be used routinely.

Table 1 and Table 2 list the experiments needed to determine repeatability using 2 approaches. For this example, the reportable value is obtained from the average of 4 individual results.

Individual Value	Reportable Value	Repeatability Standard Deviation Calculation for the Reportable Value
Value 1		
Value 2	Average 1	
Value 3		
Value 4		
Value 1		
Value 2	Average 2	
Value 3		Standard deviation is calculated as standard deviation for the 7 "averages" or "reportable values".
Value 4		
Value 1		
Value 2	Average 3	
Value 3		
Value 4		
Value 1		
Value 2	Average 4	
Value 3		
Value 4		

Individual Value	Reportable Value	Repeatability Standard Deviation Calculation for the Reportable Value
Value 1		
Value 2	Average	
Value 3	5	
Value 4		
Value 1		
Value 2	Average	
Value 3	6	
Value 4		
Value 1		
Value 2	Average	
Value 3	7	
Value 4		
	Calculate the standard deviation of the 7 reportable values.	

Table 1 Experimental Approach 1: The experimental plan is to determine repeatability by performing all the tests needed to obtain 7 reportable values. This plan requires determination of 28 individual values.

Individual Value	Repeatability Standard Deviation Calculation for the Reportable Value
Value 1	
Value 2	
Value 3	Calculate the standard deviation (s) of the 7 individual Values. Then calculate the standard deviation for the reportable value as $s/\sqrt{4}$
Value 4	
Value 5	
Value 6	
Value 7	
Calculate s, the standard deviation of 7 individual values.	

Table 2 Experimental Approach 2: The experimental plan is to determine repeatability by determining the standard deviation (s) for 7 individual values. The standard deviation of the reportable value, which is the average of 4 individual values, is the standard deviation of the 7 individual values divided by $\sqrt{4}$.

Sampling

The collection of a sample is a step that adds to the uncertainty, u, of a result. The sampling uncertainty must be considered. ISO/IEC 17025 requires that if a laboratory takes the samples, the laboratory must include the uncertainty from sampling in the final measurement uncertainty quoted for the method.

The sampling uncertainty (u_s) and the analytical uncertainty (u_a) are independent and are combined as their variances.

$$u^2 = u_s^2 + u_a^2$$
Equation 2

If the sampling uncertainty has been estimated, the analytical uncertainty can be determined by a rearrangement of equation 2 and vice versa. This can be an iterative process as the uncertainties are estimated.

Sampling is a discipline in itself and will not be discussed in detail. The reader is referred to the many specific references about sampling that are available such as the Eurachem guide *Measurement Uncertainty Arising from Sampling*[iii].

Decision rule approach and target uncertainty

Measurement uncertainty

The analytical parameters that are included in a method validation, such as accuracy and precision, are well established and understood; there is a wealth of information about these characteristics. Less understood and becoming increasingly important to analytical chemistry is the use of measurement uncertainty. Measurement uncertainty is a critical parameter that must be an outcome of method development and confirmed in method validation. Measurement uncertainty is required to demonstrate a method is fit for purpose.

The uncertainty of an analytical result is defined according to the *International vocabulary of metrology – Basic and general concepts* and associated terms (VIM)[iv] as "A non-negative parameter characterizing the dispersion of the quantity values being attributed to a measurand, based on the information used (VIM 2.26)". The term "measurand" can refer to the result of a pH measurement, a conductivity measurement, a density measurement, the concentration of an analyte or any of a number of other results of a measurement process. In each of these cases, the value obtained for the measurand can be subject to many sources of variation that result in a range of values within which the measured value can be expected to be found with a certain level of confidence. This range is the measurement uncertainty. Only when the bias and uncertainty of a measurement process have been estimated can the laboratory and/or client have confidence that the quality of the result is acceptable.

Uncertainty and error are not the same thing. The error is the difference between a single measurement result and the "true value" whereas uncertainty is a range. The error can be used to correct a result but the uncertainty cannot.

The procedure for estimating measurement uncertainty is described in the EURACHEM/CITAC guide *Quantifying Uncertainty in Analytical Measurement*[v]. Another source is the *VAM Project 3.2.1 Development and Harmonisation of Measurement Uncertainty Principles Part (d): Protocol for uncertainty evaluation from validation data* [vi].

Decision rules
A decision rule consists of a specification of the acceptance or rejection criteria of a product based on the result of a measurement. It must take into account the uncertainty estimate of the measurement and the acceptable risk of making an incorrect decision of acceptance or rejection. In order to assist in making the decision, acceptance and rejection zones are constructed which meet the acceptable level of risk specified by the client. The decision rule states that if the product lies within the acceptance zone then it is compliant. The client should be aware that the risk of accepting a

product as being compliant when it actually isn't is not zero. Similarly, the risk of rejecting a product when it is actually acceptable is not zero. The client should understand and be willing to accept the level of this risk as built into the decision rule.

The laboratory must ensure that the acceptable level of risk and the associated acceptance/rejection criteria are understood by the client when establishing specifications for meeting regulations or product acceptability. These should be part of the analytical requirement agreed on in discussions between the laboratory and its client.

Once the client has understood the concept of an acceptable level of risk of a wrong decision, discussions can begin on establishing a decision rule for the specific analytical work to be done. With the decision rule decided on the laboratory can calculate an acceptable target measurement uncertainty. When the method developed or selected is shown to be able to achieve the target uncertainty established then that method can be declared to be fit for purpose. The use of target uncertainty is described in the paper by Weitzel and Johnson [vii].

Determining the target uncertainty
In the absence of any other information a requirement for the maximum measurement uncertainty could be selected arbitrarily. For example, a relative uncertainty of 10% is chosen. With this approach the uncertainty is not related to the purpose of the method, so it could be too stringent or too lax. This approach has been taken in the past, but in view of the better understanding of the use of uncertainty in analytical chemistry, this approach is not preferred.

How to determine a target measurement uncertainty
In many circumstances, the measurement uncertainty is not known. This situation requires the determination of a target uncertainty. In order to ensure the method developed or selected meets the client's requirements the measurement uncertainty of the analytical data must be estimated.

If the laboratory is using a standard test method for which the measurement uncertainty is known, then the laboratory need only verify the method and confirm the uncertainty for the method in their laboratory is acceptable.

The customer asking for the method may not know the requirement for the measurement uncertainty and or they may expect the laboratory to know what measurement uncertainty is achievable and acceptable. Often the customer is not familiar with method development or validation and relies on the laboratory to know what to do. When the laboratory must define the target measurement uncertainty the best way is to develop a decision rule from the client's requirements and then work backwards from the decision rule to calculate the target measurement uncertainty. The process is outlined in Figure 2.

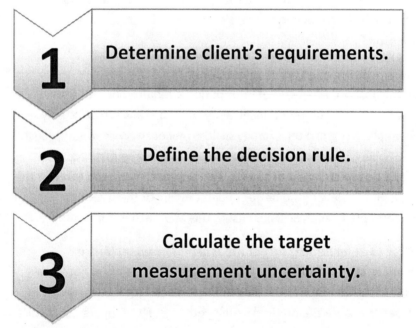

Figure 2 Process to Calculate the Target Measurement Uncertainty

Impact of Bias and Uncertainty Conundrum on Target Uncertainty

As discussed above, the characteristics of bias and uncertainty have to be considered together when assessing fitness for purpose.

If bias is shown to be negligible or is corrected for, then there is no impact on the uncertainty. If the bias is not corrected for and is significant, then an adjusted target uncertainty will have to be used. If the distribution of results is close to the limit, the range between the limit and the middle of the distribution of test results is decreased by the bias.

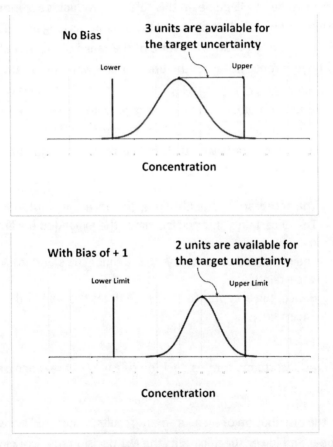

Figure 3 The impact of bias on target measurement uncertainty in an example with limits at 7 and 13 in the cases of no bias and bias.

Decision rules approach:

There are many types of decision rules. The decision rule is a form of specification that can be used to define the target MU. The basis for Decision Rules is described in the American Society of Mechanical Engineers (ASME) Standard *Guidelines for Decision rules: Considering measurement Uncertainty in Determining Conformance to Specifications*[viii].

Using the N:1 decision rule

This approach can be used when upper and lower limits are specified. It uses a simple acceptance and rejection decision rule which is described in the ASME guide. In this rule the product is accepted if the result is between the upper and lower limits as long as the N:1 ratio is met. The N:1 ratio is the ratio of the range of the acceptance zone (the difference between the upper and lower limit) to the uncertainty interval for a sample. The uncertainty interval is twice the expanded uncertainty, U, using a coverage factor, k, of 2, so the expanded uncertainty, U, is twice the standard uncertainty, u. Historically, the N:1 ratio was 10:1, but current convention has it set to 4:1. In summary:

- The acceptance zone is 4 times the uncertainty interval,
- The uncertainty interval is 2 times the expanded uncertainty, U
- The expanded uncertainty, U, is 2 times the standard uncertainty, u.
- Hence, the acceptance zone is 16 times the standard uncertainty, u.

The target uncertainty for this decision rule is 1/16 the range of the acceptance zone.

When the distribution of measurement results is centered between the upper and lower specifications the N:1 decision rule works well. In the case where the result is close to a limit, there is a higher probability that the actual value is out of specification. When the measurement process is not centered between the upper and lower specifications, then a different decision rule will have to be used. For example, a decision rule that uses the approach of "guard banding" (to be discussed below) to ensure proper decisions are made. Whether or not the material has properties that are centered will have to be established as part of the process in the selection of the decision rule.

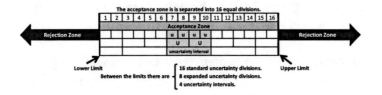

Figure 4 Relation of standard uncertainty, u, expanded uncertainty, U, and the uncertainty interval to the acceptance zone for the 4:1 Decision Rule [viii].

Simple decision rule for an upper specification limit that includes the uncertainty

When the result is being compared to an upper specification limit only, a simple decision rule is that the result is considered compliant if the result is below the limit and a result equal to or above the limit means noncompliance, provided the uncertainty is below a certain value. In this case the decision rule lists the uncertainty and the method can be developed to provide that level of uncertainty. This decision rule is based on the understanding that the uncertainty is small when compared to the limit so that the probability of making a wrong decision is acceptable.

Simple decision rule for an upper specification limit when the uncertainty is not given

When the result is being compared to an upper specification limit only, a simple decision rule is that the result is considered compliant if the result is below the limit and the result equal to or above the limit means noncompliance. In this case, the rule does not specify the maximum permitted value of the uncertainty. As a consequence, the probability of making a wrong decision is unknown. If the laboratory's client cannot supply an acceptable uncertainty, then the laboratory will have to work with the client to determine what probability of making a wrong decision is acceptable so a target uncertainty can be derived.

What does probability mean?
Most analytical data fits a normal distribution curve and this treatment assumes that. If the laboratory data is not normally distributed then the probability should be determined using the distribution function that represents the laboratory's data.

When a measurement is repeated, there can be a scattering of results about a mean value. This scatter is due to indeterminate or random errors that cannot be assigned to any definite cause. The distribution of these results can be characterized by the following observations:

- The most frequently observed value is the mean or average;
- The values are distributed symmetrically about the mean;
- There are more values closer to the mean than farther away;
- If there is no systematic error, the mean approaches the true value;
- Random error results can be treated statistically using the laws of probability.

The distribution of measurement results described is known as a normal or Gaussian distribution. Assume an infinite number of measurements was taken, the results of which were normally distributed. The frequency with which each result occurs versus the value of the result for a simulated sampling experiment is demonstrated in Figure 5. Z^{ix} is the standardized normal variable which is the number of standard deviation units from the average of the infinite number of measurements. "For a normal distribution with a mean, μ, and standard deviation, σ, the exact proportion of values which lie within any interval can be found from tables, provided that the values are first standardized so as to give Z-values. ... This is done by expressing any value of x in terms of its deviation from the mean in units of the standard deviation, σ, that is: the standardized normal variable", Miller and Miller[x]. Experimental values $\bar{x}$ and s represent the theoretical values, μ and σ, respectively.

Z is found by:

$$Z = \frac{(x - \bar{x})}{s}$$

Equation 3

For data with a standard deviation of s, the probability of finding results from that data set within a certain range of $\bar{x}$ can be predicted. These predictions are that the data will be distributed in such a way that 68.3% of the data will lie between $\bar{x} \pm Z$, 95.4% in the range $\bar{x} \pm 2Z$, and 99.7% in the range $\bar{x} \pm 3Z$. To determine the target uncertainty the area under the curve above the limit or between the limits, depending upon the decision rule, is needed.

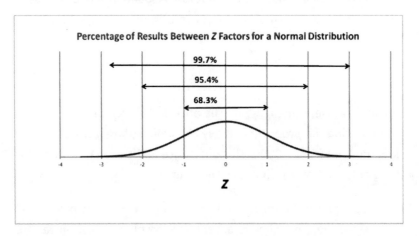

Figure 5 Percentage of results that fall between Z factors for a normal distribution.

The area under the curve can be determined using the standard normal cumulative distribution function. For example, the percentage of values above Z = 2 is 2.3%. The percentage of values below Z = -2 is 2.3%. Thus, the percentage of values between +2 and -2 standard deviations is 95.4%. Reference books contain useful tables for calculating percentage of values based on Z. However, the format of these tables is not standardized and care must be taken to ensure the tables are being used correctly.

The following procedure, to determine the percentage of values between Z factors is taken from *Juran's Quality Handbook*[xi] p. 44.33 and Appendix II, Table B. Juran's table is used here because many companies and laboratories are familiar with this reference and use it for their statistical process control.

The process of using probability to develop a decision rule for an upper specification limit when the uncertainty is not given
A decision rule is developed in order to calculate a target measurement uncertainty for a situation where an upper limit (L) should not be exceeded. To develop the decision rule the probability of getting a result above the limit needs to be known. Such a probability is found using Table B in *Juran's Quality Handbook*. The table lists the proportion of the total curve from -∞ to K, where

$$K = (X-\mu)/\sigma \qquad \text{Equation 4}$$

and X in this case, is the limit, L. The area under the entire curve is 1.0000. To find the proportion above the limit, subtract the area below the limit from 1.0000. This difference is the probability of getting an analytical result above the limit.

Two pieces of information are needed from the client; the probability of failure the client is willing to accept and the highest concentration they expect in their product. The target uncertainty is calculated so that when a sample, which is actually at the highest concentration, is analyzed several times it yields a mean result, $\bar{x}$ (which is μ in Juran's Equation 4), at the highest concentration with an acceptable probability of a single result exceeding the limit (L). Since it is necessary to take into account variation in the production process, the sampling and the analytical process, selecting the highest level experienced in the production process simplifies the development of an appropriate decision rule. This is a worst case scenario since the probability of a result exceeding the limit is actually smaller than the total probability because most production runs produce a product that is less than the maximum experienced.

The estimate of σ is s, the standard deviation. In this case it has to be determined. That becomes the target uncertainty (u_T) for the method. The formula becomes

$$u_T = \frac{(L - \bar{x})}{K}$$

Equation 5

u_T = target uncertainty
L = the limit
$\bar{x}$ = the maximum concentration the client expects in the product (this is μ above)
K is obtained from Table B in Juran, based on the probability selected by the client

Following are three examples in which the target uncertainty is derived using Table B in Juran. In all examples the limit is 100 and the client states the highest concentration the product achieves is 83.5. The laboratory works with the client to determine the probability of failure the client is willing to take. Failure occurs if a sample is taken from a lot that is actually 83.5 and an analytical result of over 100 is obtained.

In example 1 the risk the client is willing to take is 5%, in example 2 the acceptable risk is 1%, and in example 3 the risk is 0.5%. Notice that the difference in the target uncertainty when going from 1% to 0.5% probability of getting a value above K is not really a significant difference.

	Example 1	Example 2	Example 3
Limit (L)		100	
Maximum Value ($\bar{x}$)		83.5	
Percentage probability of getting a result above the limit	5%	1%	0.5%
The probability of getting a result above the limit.	(1.0000 – 0.9495 = 0.0505)	(1.0000 – 0.9901 = 0.0099)	(1.0000 – 0.9949 = 0.0051)
The probability that the value will be below the limit is the value to look up in the Table.	0.9495	0.9901	0.9949
K From Table B for the probability below K	1.65	2.36	2.52
Target uncertainty (u_T)	10	7.0	6.5

Table 3 Examples of determining the target uncertainty using probability. The probability in the table closest to that selected is used. For example, the probability of 5% or 0.0500 is not in the table, so 0.0505 is used.

This is a conservative approach towards establishing a target uncertainty, especially if large numbers of the lots are substantially below the maximum concentration. In the example above, if 1% of the lots are at that maximum value of 83.5 and a 5% probability of exceeding the limit is chosen, then there is only a 0.05% chance of obtaining a result above the limit. This probability can be used to calculate the cost of an incorrect decision when deciding upon the acceptable risk.

Setting target uncertainty when specification states the guard band

In this case the decision rule is given to the laboratory that specifies a decision rule using a guard band and the acceptable probability of nonconformance. More and more regulations are being written this way. The target uncertainty can be calculated using the decision rule to ensure the method is fit for use.

For example:

An effluent stream can be released to the environment provided the As is ≤ 10 µg/L. The decision rule states the effluent water will be considered to be compliant if the probability of the mass concentration of As in water being lower than 10 µg/L exceeds 95%. The rejection zone is set with a guard band of > 8 µg/L.

The target uncertainty that meets the three requirements in the decision rule will have to be determined. The three requirements are the ≤ 10 µg/L limit, the guard band starting at > 8 µg/L and the greater than 95% probability of conformance.

The calculation of the target uncertainty is based on the understanding that if a sample at 8 µg/L were repeatedly tested, the results would be normally distributed around 8 µg/L. The target uncertainty should be such that the probability of a second analysis yielding a result above 10 µg/L is 5%. To achieve this, the Z factor is 1.65[ix]. To calculate the target uncertainty, divide 2 µg/L by 1.65. The 2 µg/L is obtained by subtracting 8 µg/L from 10 µg/L. The target uncertainty is 1.21 µg/L.

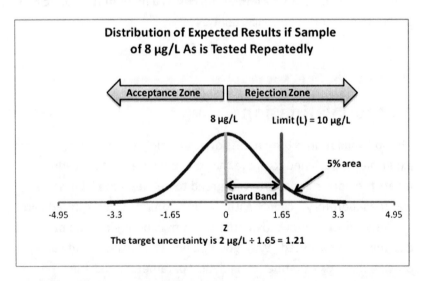

Figure 6 Distribution of expected results if a sample of 8 µg/L As were tested repeatedly showing that 5% of the results would lie above 10 µgL if the uncertainty were 1.21 µg/L.

Uncertainty and cost approach

The cost of a measurement, or of a method, is an important consideration when deciding on the required measurement uncertainty. A laboratory could develop a state of the art method with expensive, sophisticated instrumentation and perform many replicates to reduce the measurement uncertainty to the absolute minimum value obtainable with the latest technology at great expense. This expense would be wasted if the method did not need that low a measurement uncertainty.

As the measurement uncertainty gets larger, the cost of making incorrect decisions based on the test result increases. As the examples above illustrate, the probability of making an incorrect decision can be determined. Using this information, the potential cost of an incorrect decision (for example, calling product nonconforming when it is conforming) can be determined. At some point the benefit of reducing the measurement uncertainty is not warranted. The total cost is the cost of making a wrong decision and the cost of the method. The point at which the total cost is a minimum identifies the optimum measurement uncertainty.[iii]

Method validation

Method validation is the process of performing a series of experiments designed to prove that the measurement procedure meets the performance criteria agreed to between the laboratory and the laboratory's client. With the decision rule clearly stated and the target uncertainty known, the series of experiments can be designed and executed. With good planning, an experiment can provide data for a number of analytical parameters, reducing the number of experiments and the cost of validation.

Method validation pre-requirements

Method validation is part of the analytical system in a laboratory and must be supported by the laboratory procedures. The instruments, including computers and software, must be adequately qualified before being used to generate data for the validation. A procedure on how to conduct and approve method validations is needed. Also, the laboratory must have a management system, such as that described by ISO 17025, to ensure the entire laboratory system produces data that is fit for use.

Instrument qualification

Most analytical methods today use some form of instrumentation. The consistent and proper functioning of the instruments is critical to producing data that is fit for purpose. Before being placed into use, it must be demonstrated the instrument meets the supplier's specifications. A useful reference on instrument qualification is published in the United States Pharmacopeia General Chapter <1058> on Analytical Instrument Qualification[xii].

The qualification can be done by the instrument developer, the installer and/or qualified laboratory personnel. The laboratory needs to ensure the instrument is properly installed in a suitable environment. After installation, the operation of the instrument in the laboratory must be confirmed. Finally, the acceptable performance of the instrument when used for its intended use, actually analyzing samples, must be demonstrated and documented.

The specifications for the function and operation of the instrument are developed before an instrument is purchased or used. Instrument qualification should prove the instrument operates satisfactorily when running routine laboratory samples. The following aspects should be considered when setting the specifications for the instrument and qualifying the instrument:

1. The sample form, whether it is liquid or solid
2. The sample size

3. The analytical range of the instrument
    a. The maximum concentration in the matrix and in the test portion presented to the instrument
    b. The limit of detection of the instrument
4. The instrument stability
5. The interferences for the instrument
6. The precision
7. The linearity of the instrument
8. The cost of operation.

1. Sample form, whether it is liquid or sample

Ensure the instrument is capable of handling the test sample or test portion in the form it is presented. This is often straightforward, but for some sample types the final form of the sample presented to the instrument can present problems, especially in the long term. For example, the solid suspension in acid leach of soils can clog the nebulizers of atomic absorption instruments or inductively coupled plasma instruments.

2. Sample size

The instrument must accommodate an adequate sample size. This is critical when the sample size may vary or be large when the sample homogeneity is a concern. If the material being sampled is not homogenous, the subsamples and test portions may have to be larger than for homogenous materials. Conversely, if the sample is homogenous, a smaller sample could be used.

3. Range of the instrument

The instrument must be able to achieve a measuring interval that is adequate for the method. The instrument should be capable of achieving the limit of detection the instrument supplier specifies. The maximum concentration the instrument can achieve may be critical when the samples have to be diluted to fall within the range of the instrument because excessive dilution degrades the limit of detection and precision of the method.

4. Instrument stability
The instrument needs to be sufficiently stable so that drift under routine operating conditions in both the short term and long term is small enough to be acceptable.

5. Instrument interferences
Any interferences that are inherent for the instrument, such as inter-elemental interferences for inductively coupled plasma, must be demonstrated to be under control and not affect the data to the extent that the data is no longer fit for purpose.

6. Precision
The short and long term variability of the instrument must be demonstrated to match the supplier's specifications. It must also be shown to be adequate under regular operating conditions.

7. Instrument response to concentration
The instrument qualification should demonstrate there is a reproducible relationship between the instrument response and concentration. The instrument response must be able to be characterized by an appropriate algorithm such as linear or quadratic.

8. Cost of operation
The cost should be considered for routine operation so that the laboratory does not commit to using a method that becomes prohibitively expensive. This assessment should include the cost, availability and reliability of any critical supplies required to operate the instrument.

Computer and software qualification
Computers and software used for testing should be capable of complying with the specifications relevant to the testing. Commercial software, such as MS EXCEL or MS WORD, database or statistical programs, is considered to be sufficiently validated. The MS EXCEL workbooks or worksheets must be validated to ensure formulas work as expected. In-House developed software must be thoroughly documented and validated to demonstrate it works as expected. In

the pharmaceutical industry the GAMP Good Practice Guide: Validation of Laboratory Computerized Systems[xiii] describes the life cycle of laboratory computerized systems, from their start to their retirement. It is a useful guide to follow when qualifying computerized systems in the laboratory.

Management of method validation

Method validation is a planned activity that needs to be managed carefully. A procedure is needed that covers the entire process. The responsibilities for designing the validation, for performing the experiments, for reviewing and approving the data and conclusions, and for authorizing changes, all need to be assigned. The analytical parameters that need to be included in the validation must be identified. The procedure should give guidance on setting acceptance criteria linked to the purpose of the method. The process for recording data and storage of the records should be clear and documented. The procedure needs to ensure that the record of the validation, an approved validation report, documents the outcomes of the validation and includes the statement the method is fit for purpose. If the method includes sampling, then the sampling operation must be included in the validation.

The analytical procedure needs to be adequately described before it is validated. This is usually done in a method procedure. The procedure can be documented in one of many ways. It can be fully implemented in the laboratory's documentation system or it can be a draft procedure attached to the validation protocol or it can be documented in the method development report. The critical requirement is that the procedure is documented and traceable so that there is no confusion over which method was used in the validation.

The analytical procedure should include the following points.

- The procedure must be identified, either by name or number or both, so that the identification is unique and clearly understood.
- The scope of the method must be clear so the purpose or limitations of the method are clear.
- The sample type for which the validation is intended must be described in sufficient detail so that the method as validated will be applied to appropriate sample types.
- The concentration ranges or qualitative attribute that is being determined, including a detailed description of the measurand, must be stated.
- The instrumentation, equipment and reagents, including their technical specifications and grades must be described. The personnel conducting the validation need this information to adequately plan for the validation.
- The required standards, reference standards and certified reference materials must be included.
- If there are any critical environmental requirements, these must be clearly stated, including any stabilization time or holding time or shelf life of reagents or test solutions or test portions.
- The steps in the method need to be described including the handling, transporting, storage and preparation of test items.
- Any quality control checks to be done before the actual test is performed need to be described along with the acceptance criteria for these checks and what to do if a check fails.
- To ensure the equipment is working properly before it is used, the procedure must describe appropriate checks.
- The procedure must describe how the data itself is to be collected and recorded, including any calculations, and how it is to be reported.

- The criteria for approving or rejecting the test method results must be clear.
- Finally, to the extent possible, the expected uncertainty for the procedure should be given.

Ideally, a method is validated once it is completely developed and fully understood. In reality, even a well characterized method may not behave as expected and changes may have to be (or often need to be) made during the validation. The change may be minor and can be readily handled through the validation and explained in the validation records. Alternatively, the change to the method could be major. In this case the laboratory needs to assess if the validation needs to be ended and a new validation started. In some cases, the method may be found not to work and has to be abandoned and a different method developed. In all cases, any changes in requirements must be approved and authorized.

The timing of the method validation relative to the method development can vary. The method can be completely developed and a formal report approved before the method validation begins. On the other hand the method validation may be done concurrently with method development. The risk in this case is that the method may change during development. This would invalidate the completed validation activities, and require the validation to be revised and redone. If the data that is gathered during method development is documented adequately to meet quality assurance requirements, that data can be used to support the method validation. The laboratory procedure should state when the use of development data is acceptable.

With careful planning the number of experiments can be minimized resulting in many benefits such as reducing the laboratory work, speeding up the validation, and reducing costs, especially when expensive reagents are used. Using a design of experiments approach, especially for intermediate precision or ruggedness, can be highly effective at minimizing the number of experiments needed and maximizing the data generated from those that are performed. An important approach to the design of the validation is to ensure that experiments are done independently so that as many variables are varied as is practical. This approach increases the number of replicates available for statistical analysis, hence, increasing the power of the statistical tests. For example, current practice is to perform a repeatability experiment once with ten replicates. If each run in the intermediate precision experiment is executed independently, the data from several runs can be used to calculate the repeatability standard deviation and the degrees of freedom will be higher than when one run is used.

In the previous paragraph "independently" means that the portions presented to the method during the validation experiments have as little in common as possible. For repeatability the test portions and test solutions are independent. For intermediate precision each run uses a new calibration standard. Some laboratories will shut an instrument off between experiments so that each experiment is done with an instrument that has been re-set. This independence ensures as many conditions are varied during the validation as possible so the precision reflects conditions that will be found during routine use.

The sample matrix must be clearly defined. For some industries this is not difficult because the matrix is well understood, for example a drug substance in the pharmaceutical industry is a pure compound. For other industries the sample matrix is highly varied, for example, in the food industry the fat content of food can vary greatly between types of food. In the mining industry, the mineral content of the ore can vary within the ore body. If the sample matrix changes from that in the validation, the applicability of the validation needs to be assessed. This assessment must be documented. If the assessment is that the validation performed is not valid for a specific matrix then another validation must be done for that matrix. Sometimes the laboratory can use the data from some experiments in the original validation and only needs to execute a subset of the experiments in the original validation. The decision to do so must be justified and documented.

Validation is a balance between requirements, costs and what is practical to achieve. If a requirement cannot be fulfilled, the impact on the validation and suitability needs to be assessed. If the deficiency is acceptable, this deficiency needs to be stated and explained in the validation protocol, with a solid justification. If possible, a plan to collect data in the future for this analytical requirement should be implemented.

Overview of validation activities

The analytical parameters that are needed to demonstrate a method is fit for purpose have been well established. The validation consists of conducting experiments for the relevant analytical parameters to demonstrate they are fit for purpose. The analytical parameters are bias, error, trueness, accuracy, precision, (repeatability, intermediate precision, reproducibility), selectivity, specificity, identity, measurement range, ruggedness, limit of detection, limit of quantitation, sensitivity and measurement uncertainty.

The definitions and terminology are being refined to more clearly describe these parameters and how they link to fit for purpose. The terminology that is used in this book is based as much as possible on VIM3[iv] and the Eurachem Guide: *Terminology in Analytical Measurement - Introduction to VIM 3* [i].

The data is analyzed statistically and the experiments must generate sufficient data so the statistical analysis is powerful enough to assess if the method is fit for purpose. In some cases, this may not be possible and in these cases the laboratory must document the explanation for not being able to execute all needed experiments and do the best it can do. An example of this is when a certified reference material is in limited supply.

Extent of validation

The validation needs to include only those analytical parameters required to demonstrate the method is fit for purpose. For example, the limit of detection does not need to be included in the validation if a method is designed to determine the measurand at a concentration well above the limit of detection. An effective way to determine the required analytical parameters is to categorize the method as to its purpose which then identifies the associated parameters. One system for categorizing methods is used by AOAC[xiv] for method verification and another system is used by *USP General Chapter <1225> Validation of Compendial Procedures*[xv].

Validation categories

The types of methods can be divided into the categories listed in Table 4.

Category	Bias	Precision	Selectivity	Limit of Detection	Limit of Quantitation	Ruggedness	Linearity	Measuring Interval
1 Identification	No	No	Yes	No	No	No	No	No
2 Analyte at Low Concentration Quantitative	Yes	Yes	Yes	Yes	Yes	Yes	Yes	Yes
3 Analyte at Low Concentration Limit Test	No	No	Yes	Yes	No	No	No	No
4 Analyte at High Concentration Quantitative	Yes	Yes	Yes	Yes/No	Yes/No	Yes	Yes	Yes
5 Analyte at High Concentration Limit Test	Yes	Yes	Yes	No	No	No	No	No
6 Qualitative	No	No	Yes	Yes/No	No	No	No	No

Table 4 The analytical parameters that must be included in the validation for each method category are identified by "Yes". For category 4, the LOD and LOQ need be included only if the range of the method is close to the LOD or LOQ.

Category 1: Identification
The identification category of tests is used to confirm the identity of the analyte. Is the analyte what it is supposed to be?

Identifying the material is not to be confused with assessing its purity or determining its assay/strength/purity, although these are closely related.

The term identity has different meanings that are all similar and it is worthwhile to be aware of the subtle differences so as not to confuse the meanings. In the pharmaceutical industry "Identity" is a category of test and refers to a collection of tests, which taken together, confirm the identity of the material. This collection of tests preferably uses independent techniques. In the Eurachem guide on method validation [xvi] identity is an analytical parameter and confirmation of identity is the demonstration that the signal produced when the sample is measured is only due to the analyte and is not coming from the presence of a similar chemical or physical entity. This is closely related to selectivity which is the assessment of the reliability of the measurement in the presence of interferences.

The approach taken to method validation in this book considers selectivity to be an analytical parameter whereas identity is not.

Category 2: Analyte at low concentration, quantitative
The range of concentrations includes the limit of quantitation for tests in this category. If the concentration range is large, this category of tests is similar to category 4. Examples of the use of this category of tests are to quantify impurities in drug products, contaminants in the environment, and nutrients in food.

Category 3: Analyte at low concentration, limit test
This is a pass/fail or Go/No Go test in which the actual concentration is not critical. The purpose of the test is only to determine if the measurand is present above or below a specified concentration.

Category 4: Analyte at high concentration, quantitative
For tests in this category the analyte is present at a concentration well above the limit of quantitation. If the range of the analyte is close to the limit of quantitation, this category test is the same as category 2.

Category 5: Analyte at high concentration, limit test
The purpose for this test is to ensure the analyte is above or below a specified concentration and that concentration is well above the LOQ. Examples are an excipient added to a drug substance or a stabilizer added to a food.

Category 6: Qualitative
A qualitative analysis is a yes/no or detected/not detected determination using a chemical or physical test. It is important that the qualitative test is able to identify the presence of the analyte in the presence of potential interferences (selectivity). LOD is important when the concentration threshold of the test is important to ensure the result is fit for purpose.

Validation tools
The experiments that comprise a method validation use many tools to obtain the data for the different analytical parameters. Sometimes a tool may not be available, in which case the laboratory is expected to document the inability to use a tool and to do the best it can do, with justification.

Reference materials
Certified reference materials may be the most useful tool because they are used in demonstrating the bias, precision and the contribution of long term variability to the uncertainty of the method. A certified reference material is required to report the uncertainty of the reference value so that the uncertainty of the bias can be estimated by the laboratory. The ISO Guides[xvii] 31, 32, 33, 34 and 35 are useful references for understanding reference materials.

In absence of a certified reference material the laboratory can use reference materials or reference standards (different standard producers use different terminology). Material used in proficiency testing or subjected to inter-laboratory round robins can be used. For such materials the laboratory must obtain as much information as possible from the supplier of the material to demonstrate its use as a standard is justified.

Frequently there is no certified reference material that matches the matrix of a laboratory's routine samples. To overcome this, a laboratory can develop its own internal or in-house reference standards that match the matrix of routine samples. The ISO Guides are useful in identifying the steps necessary to prepare, characterize, and use in-house reference materials. ISO/CD Guide 80 *Guidance for in-house Production of Reference Materials for Quality Control (QCMs)* is being written to give guidance specifically for producing in-house reference materials.

The laboratory may also use material that has been analyzed previously, as long as the laboratory evaluates the material suitability and justifies the use of the material. The previous analyses may have been obtained by the same laboratory at another time, or by another laboratory, or they may have been obtained by using a different method. Again, the suitability of that data must be assessed and justified, especially taking into consideration the uncertainty of the value assigned to the material from the previous analyses.

When the reference material is not identical to the sample, the laboratory may justify the use of a reference material with a different matrix. This is acceptable if the laboratory can demonstrate that the different matrix does not have a significant effect on the analytical result.

If no matrix matched reference material is available the laboratory may choose to use spikes, where a known amount of analyte is added to the sample matrix. The shortcoming of spikes is that they may not adequately represent the analyte, or its behavior in the actual sample. The material used for spiking may not be tied to the matrix in the same way as the actual analyte. This is especially the case when a digestion is used and the spiked material is readily available and does not have to be "leached" from the matrix as the actual analyte does. In the absence of suitable reference materials, the laboratory must do the best it can and sometimes using spikes is the only tool available.

Blanks

An important part of a method validation is demonstrating that nothing is detected when nothing is there. In reality a true blank is difficult to achieve. Blanks can be introduced at many levels such as samples that do not contain the analyte or reagent blanks that contain only reagents. Blanks can be introduced at different steps in the method such as at the sample preparation step or at the test solution step.

The data from the blank is used to demonstrate selectivity. The data may also be used to calculate the reportable value, in that the value of the blank is subtracted from the value of the sample. The blank may be used as a quality control monitor to demonstrate there is no contamination.

Interference checks

In order to demonstrate selectivity, the laboratory needs the potentially interfering material. This could be chemicals that are known to be in the sample, or potential contaminants which may or may not be in every sample, contaminants from the method reagents, or environmental contaminants. The laboratory is required to assess all the possible interferences, decide which to include in the validation using probability and risk analysis, and to demonstrate the possible interferences are not significant. If avoidance of interference is not possible, the result may require correction for that interference.

Replicates, especially duplicates
Using replicates can be a valuable tool for demonstrating precision. The replicates can be created at any step to demonstrate the variability for different steps of the method. The number of replicates inserted should be sufficient to ensure a reliable statistical analysis. Duplicates are a special type of replicate that can generate data for calculating the standard deviation for the method that includes the variability introduced from the step in which the duplicate is inserted to the end of the analysis.

It is important that the replication be done consistently so that each replicate includes the same variables.

An important use of replicates is to establish precision when a reference material in the same form as the sample is not available. This is often the case when the laboratory sample must be reduced in form and size such as grinding grain or crushing rock and the reference material is available in powdered form only. The inclusion of replicates, especially duplicates, at the size reduction step will yield data on the variability from that step on. This variability can be combined with that from the reference material to yield data for the precision of the entire method.

Duplicates
Given enough duplicate data, a reliable estimate of the standard deviation of the measurement system, and thus its precision, can be calculated. This is accomplished with the use of a simple calculation as follows:

$$Variance = v = s^2 = \frac{\sum R^2}{2N} \qquad \text{Equation 6}$$

R = the difference between duplicates
N = the number of duplicate pairs

There are three concerns that need to be addressed to ensure that the estimate of the standard deviation "s" calculated with this method is reliable:

1) N should be at least 20.

2) The concentrations of the samples from which the duplicates were selected should not differ by orders of magnitude (10 ppm to 500 ppm for example, is probably too large a concentration range - try calculating the variance for two ranges in this case: <100 ppm and > 100 to 500 ppm).

3) The duplicates must not exhibit a bias between the duplicate sets - the sum of the differences between the individual duplicate pairs (with the sign taken into account) must be close to zero.

There are three advantages associated with using duplicates to estimate the precision of a process. First, it provides a realistic estimate of the precision since it is based on actual samples. Repeated measurements on a reference sample may not give the same level of variability that is found with real samples going through the process. Second, the overall variance obtained reflects the sum of the variances introduced at each step of the process after the inclusion of the duplicates. Thirdly, the standard deviation obtained from this data includes sampling uncertainty - the variance due to differences between duplicates.

Design of experiments and statistics
Design of experiments (DOE) and statistics are powerful tools for method validation. They allow an initial estimate to be made of the measurement uncertainty of the method and this is required to demonstrate the method is fit for purpose, as was discussed earlier. Also, if used correctly, they can reduce the number of experiments required thus making the validation more efficient and effective.

The method validation can be planned so that one experiment can fulfill multiple purposes. For example, the experiment to determine the bias can also yield data for the repeatability. The planning should ensure the experiments are executed independently so that as many variables as possible are varied, with little extra work, to ensure an initial estimate of the measurement uncertainty with as many degrees of freedom as possible is obtained. Uncertainty estimated at the validation stage can only be considered as an initial estimate. An estimate of the measurement uncertainty for the long term will have to be based on data collected over a period of months. The uncertainty must be continually verified during the life of the method.

Design of experiments allows the laboratory to gain maximum benefit with minimal testing in its method validations. A description of a design of experiments for ruggedness[xviii] is given below in which the impact of 7 variables is determined with 8 experiments.

Statistics is becoming an increasingly important tool for method validation. In order to use statistics correctly and to take advantage of its usefulness and power the laboratory requires an understanding of basic statistics. One such tool is the analysis of variance (ANOVA). When the experiments are planned correctly, the ANOVA function in MS EXCEL can be used to isolate the different precisions such as repeatability and intermediate precision. Using ANOVA instead of estimating repeatability from one experiment with 5 degrees of freedom, the validation can provide an estimate from 5 experiments with 18 degrees of freedom.

Statistical tests such as t tests are useful for demonstrating statistical significance. However, a difference may be statistically significant and not be practically significant. This should be included in the acceptance criteria of the validation. The target measurement uncertainty makes the assessment of practical significance straightforward because the method performance is compared to the target uncertainty to show it is fit for purpose. For example, the between-day variability may be shown to be statistically significant by the F test. However, as long as the precision for either day does not cause the overall uncertainty to be above the target uncertainty, the difference is not practically significant.

MS EXCEL as a validation tool

MS EXCEL can be a valuable tool for method validation. A template identifying the data that is needed and containing the formulas can be prepared, checked to ensure it is correct, and then used as needed.

Validation activities

Validation activities are the experiments that need to be run to demonstrate and prove a method's capabilities. The experiments can be examined and organized according to the analytical parameter being examined. For each parameter the purpose is given, the experiments are described, the acceptance criteria are suggested and finally the parameter is discussed.

Selectivity/specificity

Purpose

The term selectivity is currently preferred over specificity with the latter not being recommended by IUPAC[xix]. Also, specificity is not defined in the Eurachem guide to terminology [i]. Selectivity is the capability of the method being validated to provide a measured value or result for an analyte that is not affected by other constituents of the material being analyzed.

Selectivity may need to differentiate between two or more forms of a constituent. These could be different oxidation states, different chiral forms, or the analyte being complexed to different ligands.

Analytical activity

If there is the possibility of a potentially interfering constituent being present then the capability of the method to discriminate between the analyte and the interfering constituent can be assessed by spiking a sample of known analyte concentration with the potential interferent and determining if there has been any difference in the analyte value. Alternatively a blank can be spiked with the interferent and a test run to see if there is a false positive value found for the analyte.

If it is not known if an interferent is present, the results from the method being validated can be compared to results from other methods or techniques. Preferably these other methods are based on alternate detection mechanisms or fundamental properties.

Acceptance criteria
Interferences can affect the bias, precision or both. As long as the analytical results meet the target uncertainty and bias requirements, then the interferences can be tolerated. The limit of detection may also be impacted, but as long as the concentration determined for the limit of detection meets the acceptance criteria it is acceptable.

For techniques such as HPLC, the USP gives guidance in the Medicines Compendium[xx] on acceptance criteria for resolution to demonstrate selectivity.

Discussion
A lack of selectivity may be caused by a contaminant, an interfering constituent, instrumental noise or some other source. The laboratory is faced with the challenge of deciding which potential source might result in a lack of selectivity and how likely it is that the source is present. It is not possible to cover off all possibilities but the most likely ones should be identified and their interference assessed. As mentioned above, in the presence of interferences the limit of detection can be affected as can the precision and bias of the method. Method validation would have to include such effects and their influence on the reliability of measurement results obtained when such interferences occur.

Selectivity is often not a problem. Even so, it is important that the laboratory not take this for granted. Method validation should be able to prove the method delivers data that is fit for purpose for all samples likely to be encountered. This will prove to be more difficult as the number of different sample types and the variety and concentration ranges of sample constituents increase. The validation planning will have to accommodate such situations.

Limits of detection and quantitation

Purpose

Limit of detection
The purpose of the limit of detection of a method is to establish the lowest concentration at which the measurand can be shown to be present, with a specified level of confidence, usually 95%, using that method. The uncertainty of the measurement around the limit of detection approaches the size of the measurement itself (100% Relative Standard Deviation, *RSD*).

For qualitative tests the limit of detection is defined as "The lowest amount of concentration of the analyte in a sample, which can be reliably detected (but not necessarily quantified). LOD_{50} is defined as the level at which 50% of the replicates are positive, and may be used as a measure of the limit of detection."[xxi]

Limit of quantitation
The purpose of the limit of quantitation of a method is to establish the lowest concentration at which the measurand can be established with a known level of uncertainty and trueness. Measurand values close to the limit of quantitation value should not be used for decision making since the uncertainty at this concentration is still large compared to that value. The limit of quantitation is not defined in VIM 3.[iv]

Analytical activity

Limit of detection - quantitative

Make one measurement on each of 10 independent sample blanks or on each of 10 independent sample blanks fortified at the lowest acceptable concentration, which is less than 10 times the expected concentration of the limit of detection. The sample blanks must be separate preparations that are matrix matched to samples as closely as possible. The standard deviation "s" in each case is multiplied by 3 and the limit of detection is calculated as the mean of the sample blanks + 3s. If the mean is negative, it is calculated as 0 + 3s.

These determinations can be used in the validation for repeatability at this concentration level since they have been run under repeatability conditions.

Limit of detection – qualitative

The method precision of qualitative tests is generally expressed as false positive/false negative rates and is determined at several concentration levels. This can be done by making randomized analyses of 10 independent replicates for each set of blanks spiked at several different concentrations of measurand. In the example found in Table 5 it can be seen that a positive result is not attained with 100% reliability for less than 100 mg/kg. This is only one example of determining a limit of detection in a qualitative test.

Concentration mg/kg	Number of replicates	Positive/negative results
500	10	10/0
300	10	10/0
100	10	10/0
75	10	6/4
50	10	2/8

Table 5 An example of establishing the limit of detection for a qualitative analysis.

Another example is the probability of detection (POD) adopted by the AOAC for qualitative microbiological work determining the Colony Forming Units (CFU): "The probability of detection (POD) function p(d) is the functional relationship between the probability p of obtaining a measurement result 1, and the contamination of the test material, quantitatively expressed as the number of CFUs of the microorganism of the defined type per unit of weight or volume, d (CFU/g or CFU/mL)"[xxii].

Limit of quantitation

One means of estimating the limit of quantitation is to use the standard deviation s calculated from the repeatability blank analyses used to calculate the limit of detection (above) and multiplying s by a factor of 10 (although factors of 5 and 6 have been used by some). A more robust approach is to spike, in replicate (3 or 4) and at different concentrations close to the anticipated limit of quantitation, several aliquots of the blank. The blank should be matrix matched as closely as possible to the anticipated or known sample matrix. Measure each set of replicated spiked blanks and calculate the standard deviation for each. Plot the relative standard deviation from each set against the spiked concentration of each set. Then the limit of quantitation is the concentration at which the relative standard deviation is 10%.

Since these analyses are run under repeatability conditions, they can also be used to validate repeatability at this concentration range.

Acceptance criteria

The actual concentrations required for the limit of detection and limit of quantitation are established as part of the method requirements. The values obtained should be equal or less than those required.

Discussion

The calculation of the limits of detection and quantification using the standard deviation are commonly used today. However, the definitions of limit of detection and limit of quantitation are still being debated. There is no agreement as to what factor the standard deviation of the blank should be multiplied by to estimate a value for either. In addition, even the name "limit of detection" or "detection limit" is not agreed on. VIM 3 [iv] refers to it in terms of minimum "measured quantity value" while IUPAC prefers minimum "true quantity value".

The laboratory should specify the convention used in calculating the limits of detection and quantitation. Many commercial laboratories quote the limit of quantitation as their limit of detection as a precaution to reporting a measurand being present when it is not. This is bad practice and should be discouraged. The limits of detection and quantitation are related to the performance of the method.

Measuring interval (range)

Purpose
Any validation of a method must include a determination of the concentration range over which a method will produce data that is fit for purpose. This is the concentration range for which the bias, uncertainty and measurement or instrument response is acceptable. Evaluation of the measuring interval is required to define the calibration protocols that will be used for the method. This would include the range of concentrations over which a single point calibration can be used for example. It will also give guidance for deciding which samples will have to be diluted or pre-concentrated in order to enable the method to accommodate the measurand concentrations anticipated to be found in the samples.

Analytical activity
Since bias, uncertainty and measurement or instrument response are all needed to define the measuring interval there are several ways to structure the analytical activity. Depending upon how the

experiments are structured, the same experiment will yield the data for bias, repeatability precision, and measuring interval.

The bias and uncertainty data are taken from the respective experiments. The measurement or instrument response is assessed separately and the three assessed together to demonstrate the measuring interval.

Use of reference samples or spikes
If at all possible, certified, or at least well characterized, reference samples that cover the range should be used to assess bias and uncertainty of the method across that range. At least 6 reference samples evenly spread over the concentrations included in the range should be run under repeatability conditions. A minimum of 5 independent replicates of each reference sample should be run. Independent replicates represent separate test portions put through the measurement process. If such a set of reference samples is not available, then spiked blanks can be used. In this case ensure that the matrix of the test solutions will match that of the samples to be run as closely as possible.

Instrument response
With this approach the response of the instrument to the analyte alone, without the sample matrix is assessed to determine the concentrations for which the instrument response is acceptable. This approach does not include bias or uncertainty; separate experiments are done to determine bias and uncertainty in the actual sample matrix. At least 6 test portions or test solutions evenly spread over the concentrations should be run once.

Linear interval
If it is appropriate, such as when a single point calibration will be used, the same set of analyses can be used to determine the linear interval which is that range of concentrations over which the measurement response is a linear function of changes in concentration. To assess linearity, visually examine the plot of the

response on the y axis and concentration of the measurand on the x axis to identify the approximate linear interval. Plot the residual values (difference between the actual y value and the y value predicted) for each x value. The plot should show a random distribution about the straight line. If the plot does not show random distribution, assess if the divergence from linearity makes the data not fit for use. Record the correlation coefficient, y-intercept, slope of the regression line and residual sum of squares so they can be used to establish acceptance criteria for the calibration line during routine use.

Acceptance criteria

The acceptance criteria for the measuring interval are those criteria for bias and uncertainty which show the data is fit for purpose. The bias and uncertainty established as the original requirements for the method are the acceptance criteria required of the measuring interval.

Discussion

Phrases used when discussing the concentrations over which a method produces data that is fit for purpose include "measuring interval", "range", linear range", "working range", and "measurement range". The terms used here are measuring interval, range and linear interval.

There is a distinction made between the range and the measuring interval. The measuring interval is defined by the values at the lowest concentration and the highest concentration for which the method gives acceptable results. The range is the difference between these two values. As an example, if the limit of quantitation for a method is 0.1 µg/mL and the upper limit is 10 µg/mL then the measuring interval is given as [0.1, 10] µg/mL. The range is referred to as the difference between the two limits of the measuring interval and in this example is 9.9 µg/mL.

The relationship between the concentration and the analytical response does not have to be linear; it has to be well characterized and repeatable. For example, the relationship could be quadratic and meet these requirements. A highly curved relationship, however, can result in a significant loss of precision. A regression analysis alone will not confirm linearity of response. In a linear regression, the correlation coefficient, r, is not an indication that the relationship is linear, it is an indication of how good the relationship between concentration and response is, be it linear or quadratic or whatever.

The acceptance criteria are stated in terms of the lower and upper concentrations over which the method must produce results that are fit for purpose. If a linear interval is also required, such as when a single point calibration will be used, that interval must be stated as part of the method validation acceptance criteria.

The concentrations in the measuring interval are expressed in the form in which the sample is presented at the measurement step – a solution aspirated into an instrument for example. The laboratory should be clear in the description of the measurand whether it is relevant to the actual laboratory sample, as given to the lab, or is relevant to the sample which is presented to the measurement step, the test portion or test solution. The validation must include any steps to convert the laboratory sample to the test portion or test solution presented to the measurement step of the method.

Accuracy, trueness and bias

Purpose
The concept of accuracy is used to compare the result from a single measurement with an accepted value (for example, the reference value of a certified reference material). Accuracy includes both systematic and random components of error - that is trueness and precision. As such, it cannot be used to give a quantitative assessment of the value of an analytical result. Results are said to be

"more accurate" when errors, including either or both the trueness and precision, are reduced. This is illustrated in Figure 7.

Measurement trueness is the "closeness of agreement between the average of an infinite number of replicate measured quantity values and a reference quantity value (VIM 2.14)"[i]. Trueness is a hypothetical concept and is not expressed numerically since it is not a quantity. Trueness is however, related to systematic error which may be expressed as a difference between the mean of numerous measurements of a certified reference value and the reference quantity value of the certified reference material. This difference is known as the bias. Figure 7 patterned after the Eurachem Guide to Terminology [i] and the AMC Technical Brief on terminology[xxiii] illustrates the relationship between error, bias, trueness, accuracy and precision.

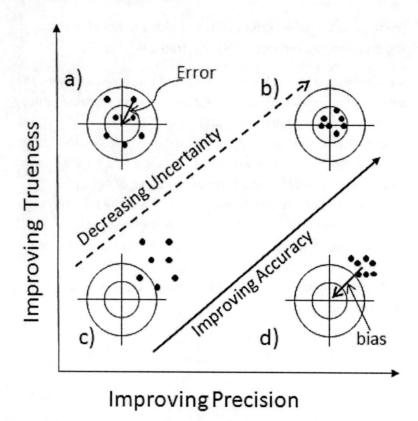

Figure 7 Each dot on the target represents an individual measurement result. The center of the target represents the reference value (usually of a certified reference material). Since accuracy includes both trueness and precision, case b) represents the best accuracy. The accuracy is poorest for case c) because there is a bias and the precision is poorer than in b) and d). Error is the difference between one result and the reference value. Bias is the difference between the mean of several results and the reference value.

Analytical activity

The analytical activity is the demonstration of the bias. To do this, analyze a certified reference material using 10 different test portions taken from the reference material along with 10 reagent blanks. Subtract the mean of the reagent blank readings from the mean of the 10 reference material results to give a mean blank corrected value. The difference between this mean blank corrected value and the reference value is the bias. If a satisfactory reference material is not available (not an unusual situation) validation could be done by using a well characterized material that is similar to the samples, such as an in-house reference material. Another alternative includes using spikes, but the possibility of the spike not behaving the same as the analyte must be considered. A third option is to compare the results against those from an alternative method that is validated and demonstrates an acceptable level of uncertainty.

The bias should be evaluated across the concentrations specified by the measuring interval of the method. This would constitute a minimum of 3 reference samples at low, medium and high concentrations, each run as 10 replicates.

In the case of spiking the recovery is calculated as:

$$\text{Recovery}(\%) = 100\% * \frac{(c_1 - c_2)}{c_3} \qquad \text{Equation 7}$$

where

c_1 = result for spiked sample
c_2 = result for un-spiked sample
c_3 = Spike concentration/amount

If planned correctly, the experiments to estimate bias can be used to assess precision at different concentrations as well. This can limit the number of experiments required for validation.

The bias is estimated by subtracting the reference value found on the certificate (X_{RV}) from the value ($\overline{X}_r$) the laboratory obtained for the reference material in the repeatability study.

$$\text{Bias} = \overline{X}_r - X_{RV} \qquad \text{Equation 8}$$

The uncertainty of the bias (s_b) is estimated by combining the uncertainty from the laboratory's estimate of the bias with the uncertainty of the reference value of the reference material (see Equation 11 below). The former is estimated using the repeatability standard deviation and dividing it the by the square root of the number of determinations, $\sqrt{n}$. The latter is obtained from the certificate of analysis of the reference material.

The uncertainty of the bias is used to determine if the bias is significant. This is important because an insignificant bias need not be of concern. In order to determine whether or not the bias is significant, establish if the range of the bias includes zero (Bias $\pm$ ts_b includes 0). If it does, the bias is not significant. The student's t value will depend on the number of degrees of freedom in the estimation of s_b and the confidence level (a 95% confidence level is the usual level for certified values in a reference material.) If 0 is close to either end of the range, it may indicate there is a bias and the laboratory may want to investigate further.

Acceptance criteria
As long as the analytical results meet the target uncertainty and bias requirements, then the bias is acceptable.

Discussion
Accuracy encompasses both random and systematic error of measurement as stated above. Measurement uncertainty is becoming a more frequently used expression of accuracy. Bias on the other hand is the result of systematic error. Error of measurement is the difference between a measurement result and the true value. In practice, an accepted value is used since a true value cannot be known. Random error is the result of unpredictable variance that is found in repeated tests of the same measurand. A systematic error of measurement is "A component of the error which, in the course of a number of test results for some characteristic, remains constant or varies in a predictable way." [xxiii]

In a method validation only bias is demonstrated.

Term	What it is	Comment
Error	Result minus the reference value	The true error is unknowable because the true value is unknowable. Since the true value cannot be known, a conventional value, such as the reference value for a certified reference material can be used giving a practical value for the error.
Bias	Systematic error	Bias can be estimated by the difference of the mean value of several measurements from the reference value. Bias can be estimated measuring the value of one or more reference materials several times under repeatability or intermediate precision conditions and calculating the mean. The difference between the mean and the reference value is the bias. In a method validation, bias is the only concept in this group that is validated.
Trueness	Closeness of agreement between the average of an infinite number of results and a reference value	Trueness is a hypothetical indication of the ability of the method to yield results close to the expected reference value. It is hypothetical because an infinite number of results cannot be obtained and the true value cannot be known.
Accuracy	Closeness of agreement between result and a true value	Accuracy describes how close a single result is to the true value. Therefore, accuracy includes the systematic error and the random error that impacts that result. Since the true value is not known, accuracy cannot be given a numerical value, but is a descriptive and comparative term for a method. A method that has less random error, or a smaller bias, or both is called "more accurate".

Figure 8 Descriptions of terms used when discussing accuracy.

A common source of systemic error is incomplete recovery. Recovery is a measure of the degree to which a measurand has been extracted from a host matrix and been measured. It gives the laboratory a tool to monitor the effectiveness of an extraction or separation step in a measurement process. This may be of interest if only one species or oxidation state of a measurand is of interest. Conversely, the laboratory may want to know if the measurand is not completely extracted by a digestion technique for example or if it is being lost through evaporation, precipitation, complexation with a ligand or from some other phenomenon.

Recoveries are usually reported as the percent of the known value that is obtained by the analysis. Thus, if a spike of 100 ng is added and the analysis result was 80 ng found then the recovery is reported as 80%. The levels of recovery can vary enormously, especially for measurands such as pesticide residues and drug constituents, for example, and it is not unusual to find recoveries ranging from 60% to 120% being acceptable in such situations.

Precision (repeatability, intermediate precision and reproducibility)

Purpose
Precision is a numerical expression of the spread of repeated measurements. This value is an important indication of the reliability of the results of measurements made using a method. Precision is expressed as a standard deviation or sometimes as a relative standard deviation.

Repeatability
Repeatability precision provides information on the spread of results resulting from a series of measurements involving the same analyst using the same method and the same instrument over a short period of time – usually in the same run. These conditions are known as repeatability conditions.

Intermediate precision

Intermediate precision is the short term precision in one laboratory. The conditions consist of different analysts performing the analysis on test portions taken from the same laboratory sample on different days. They may or may not be using different instruments. It is possible other conditions are varied. These replication conditions give intermediate precision. The conditions that are varied should be described when the intermediate precision is stated.

Reproducibility

Reproducibility is the precision between different laboratories. When different analysts perform the same method on the same sample in different laboratories using different instruments on different days, the precision resulting is known as reproducibility. This is what is achieved in round robins and proficiency testing for example if the same method is used.

Analytical activity

Repeatability

Repeatability precision is estimated by running, under repeatability conditions as described above, 10 replicates of samples at measurand concentrations that cover the measuring interval of the method. In the case where the measuring interval covers a large concentration range (more than two orders of magnitude), reference samples at low, medium and high concentrations should be run 10 times each. In the case of a narrow range of concentrations expected in the measuring interval it is possible that one reference standard analyzed 10 times would be sufficient. The replicates should be independent. The number of replicates given here is taken from the Eurachem Guide on method validation.[xvi] Different standard setting bodies recommend different numbers of replicates.

Intermediate precision

Intermediate precision would be estimated in the same way as repeatability by using the intermediate precision conditions discussed above. In certain instances the same laboratory sample cannot be used, for example the laboratory sample may not be stable enough to be used on two different days, or the material may be too expensive or only a limited amount is available. In these cases the laboratory must assess the impact on the intermediate precision of using different laboratory samples.

The laboratory needs to identify the sources of variability that can be varied in the intermediate precision studies. This can be done by reviewing the test method to identify steps where variability can occur. Examples of such sources are analysts, equipment, instruments, reference materials, lots of reagents, etc. A fish bone analysis such as discussed in the Eurachem guide for estimating uncertainty [v] is useful for this review. Next, the laboratory decides which sources can and will be included in the intermediate precision study. Note that the uncertainty estimate for intermediate precision will only include those sources which vary during the study, so the intermediate precision study should be as comprehensive as possible.

Reproducibility
Reproducibility, the between laboratory precision, can be obtained through collaborative studies which are usually run by official method organizations or proficiency testing providers. This entails the careful preparation of samples to ensure homogeneity of the measurand being studied. ISO 5725-2:1994[xxiv] gives guidance on the general principles for estimating reproducibility from collaborative studies.

Use of ANOVA to estimate the various precisions
In order to separate and derive an estimate of the different sources of variation, a one-way ANOVA treatment of the data can be used[xxv]. The MS EXCEL® output shown in Figure 9 below illustrates this. The Mean Squares (*MS*) are used to calculate the standard deviations using the formulae in Figure 9.

	Condition 1	Condition 2	Condition 3	Condition 4	Condition 5
Replicate 1	10.1	10.2	9.8	10.2	10.5
Replicate 2	10.3	9.7	10	10.3	10.6
Replicate 3	9.8	10.1	10.3	10.1	10.6
Replicate 4	10.1	10	10.1	9.8	10.5
Count	4	4	4	4	4

Anova: Single Factor

ANOVA

Source of Variation	SS	df	MS	F	P-value	F crit
Between Groups	0.802	4	0.2005	5.49315068	0.00628861	3.055568
Within Groups	0.5475	15	0.0365			
Total	1.3495	19				

Precision Estimates Using ANOVA

			Formula
Repeatability	S_r	0.191	=SQRT(MS Within Groups)
Between Group Standard Deviation	S_{BG}	0.202	=SQRT((MS Between Groups - MS Within Groups)/Count)
Intermediate Precision	S_{IP}	0.278	=SQRT($S_r^2 + S_{BG}^2$)

If the condition is different days, or analysts, or instruments, etc., the formula for S_{IP} yields Intermediate Precision.

If the condition is different laboratories, the formula for S_{IP} yields Reproducibility and S_{WG} is the repeatability pooled across the laboratories.

Figure 9 MS EXCEL output used to estimate the various precisions.

Duplicates

Often it is difficult to obtain the precision for an entire measurement process. This is the case when there is a sample preparation step that reduces sample size through sub-sampling and particle size through crushing and or grinding the laboratory sample. These steps are not included in the measurement process when a reference material is used. In these cases, duplicates as discussed above can be used in the method validation.

Acceptance criteria

The acceptance criteria are linked to the target uncertainty. Precision is an important uncertainty component and may be the largest component. The validation has to prove that none of the precision components of uncertainty are so large that the target uncertainty cannot be achieved. If the uncertainty obtained from the validation is close to the target uncertainty, the laboratory may want to reduce the uncertainty and can do so by examining the precision components of uncertainty and focusing on the largest component. For example, the laboratory could increase the number of individual results used in the reportable value. The uncertainty in the reportable value would then reflect the standard error of the mean as discussed above under reportable value.

Discussion

If the experiments are designed appropriately, Analysis of Variance (ANOVA) can be used to extract the repeatability and intermediate precision values from one set of experiments, reducing the amount of testing. Also, the same experiments can be used for the bias estimation.

Ruggedness (or robustness)

Purpose

Ruggedness testing identifies the tolerance a method has towards small, uncontrolled changes in its operating conditions. A ruggedness test is an intralaboratory experimental plan to examine the behavior of an analytical process when small changes in the environmental and/or operating conditions are made, similar to those likely to be experienced when the method is used routinely.

Analytical activity
An experiment to show ruggedness can be as simple as changing one or two conditions suspected to have a significant effect on the measurement result. A design of experiment (DOE) approach can be used to more thoroughly demonstrate ruggedness as discussed below.

Ruggedness test
The following scheme allows ruggedness testing to be done with a minimum of effort that permits the testing of 7 conditions using 8 tests.

The concept is to monitor the effect of the change of one condition while any effect due to the change of another is averaged out. If it is assumed that seven basic conditions (temperature, distillation time, instrument set-up, sample to flux ratio or whatever) are assigned two levels. One level is designated **A, B, C, D, E, F,** and **G** while the alternate level is designated **a, b, c, d, e, f** and **g**. For example, a condition could be temperature and level **A** is a high temperature and level **a** is a low temperature. The temperature range is within that most likely to be experienced during routine use of the method.

Using this terminology it is possible to set up 8 tests in the combination shown in Table 6:

Condition	Run Number							
	1	2	3	4	5	6	7	8
A or a	A	A	A	A	a	a	a	a
B or b	B	B	b	b	B	B	b	b
C or c	C	c	C	c	C	c	C	c
D or d	D	D	d	d	d	d	D	D
E or e	E	e	E	e	e	E	e	E
F or f	F	f	f	F	F	f	f	F
G or g	G	g	g	G	g	G	G	g
Result	s	t	u	v	w	x	y	z

Table 6 Ruggedness Testing Combinations

Table 6 shows the levels for each condition and their alternates for seven conditions with 8 different tests being run. Four tests will be run with a condition at one level and 4 tests will be run at the alternate level. The organization of these conditions as outlined in the table allows the effect from a change of level **A** to **a** to be assessed by comparing the average (s+t+u+v)/4 with (w+x+y+z)/4. Each of the other conditions (**B** and **b** for example) occur twice in each of the two averages for **A** and **a** and therefore cancel out within each average. This leaves only the variation due to changing from level **A** to level **a** being reflected in the difference between the two averages. The same can be done for the remaining 6 conditions (compare the average (s+t+y+z)/4 with (u+v+w+x)/4 for the effect of changing from condition **D** to **d** for example).

The differences can then be ranked in decreasing order of the magnitude to determine which conditions (if any) do or do not have a significant effect on the results. Thus the ruggedness of the method is assessed for the 7 conditions tested.

Test for Significance of Differences

1) Before running the 8 tests, obtain an estimate of the within-run precision, as a standard deviation "s", from "N" replicate analyses of a representative sample.

2) Calculate the test statistic "t":

$$t = \frac{\sqrt{n} * Dx_i}{\sqrt{2} * s}$$

Equation 9

where Dx_i is the difference for parameter "i" from "n" measurements for each level for that condition (4 in this case) and "s" is the standard deviation from step 1 above.

3) Compare "t" with the 2-tailed critical value, t_{crit} for N-1 degrees of freedom at 95% confidence where N is the number of tests used to obtain "s" in step 1.

4) If t is < t_{crit} the difference is NOT significantly different from zero. If t is > t_{crit} then variations in the conditions for which this is true have an effect and must be controlled.

5) If a condition is found to be significant the laboratory can implement the method with careful control of that condition or it could improve the method.

Discussion

The terms ruggedness and robustness have been used interchangeably by many organizations over the years. Recently ruggedness has become more commonly used and is used here.

The purpose behind ruggedness is to challenge the precision of the measurement process more than is done in the experiments to demonstrate intermediate precision. The ruggedness experiments are designed to show the results of the measurement process are not significantly impacted by small changes or variations that may occur in the procedure during normal routine use.

The conditions to be varied in the ruggedness test are found in the documentation for the measurement process. They should have been identified during method development and measures taken to control any conditions that have a significant effect. These conditions are also called influence quantities and are an important aspect of traceability. A requirement of traceability is that the influence quantities are identified, are assessed for significance and are under adequate control. The ruggedness test does this and is a useful tool for demonstrating this aspect of traceability.

Design of experiments is a discipline in itself. More information on DOE and the ruggedness test presented above can be found in the AOAC book *Use of Statistics to Develop and Evaluate Analytical Methods* [xviii] and the book *An Introduction to Design of Experiments, A Simplified Approach*[xxvi], by Larry B. Barrentine.

Measurement uncertainty

Purpose
The initial estimation of measurement uncertainty as part of the validation process permits the laboratory to assess the performance of the method against the established target uncertainty. This allows the method to be assessed for fitness for purpose.

Analytical activity
The data from the validation experiments described above can be used to estimate measurement uncertainty.[xxvii] This uncertainty reflects only a short term uncertainty estimate and will have to be modified as data is collected during routine use of the method to add the long term components of uncertainty.

Several validation experiments can be used to obtain an initial estimate of uncertainty and each of these is discussed below. The table below shows the relationship between the validation experiments and the uncertainty components they cover.

Uncertainty Component	Bias Study	Repeatability	Intermediate Precision	Ruggedness
Analyst			X	X
Instrument Short Term	X	X	X	X
Analytical Manipulations	X	X	X	X
Extraction	X	X	X	X
Bias	X			

Table 7 Uncertainty components related to each validation parameter.

Bias

The uncertainty of the bias (s_b) is a combination of the uncertainty from the laboratory estimate of the bias (s_w) and the uncertainty from the reference value (s_μ).

The bias is estimated by subtracting the reference value from the value the laboratory obtained for the reference material in the repeatability study. The laboratory contribution to the bias uncertainty is calculated using Equation 10 where *n* is the number of replicates used in the repeatability study.

$$s_w = \frac{s_r}{\sqrt{n}} \qquad \text{Equation 10}$$

The uncertainty for the bias is obtained by combining these two components as shown in Equation 11.

$$s_b^2 = s_w^2 + s_\mu^2 \qquad \text{Equation 11}$$

Precision

Precision is an essential component of the overall uncertainty for a method. There are three types of precision, repeatability, intermediate precision and reproducibility, which are discussed above in the precision section. The method validation can generate multiple estimates for repeatability and intermediate precision. If the validation includes multiple laboratories, then data for reproducibility is also available. For a direct estimate of the overall uncertainty, the intermediate precision is usually all that is needed. If the laboratory needs a more detailed understanding of the method and the various uncertainty components, the laboratory can analyze the data using a technique such analysis of variance (ANOVA) as discussed above in Use of ANOVA to estimate the various precisions.

The intermediate precision study can cover most of the uncertainty arising from precision if it is designed correctly. The laboratory can identify the uncertainty components using Table 7. If all the sources of variability are challenged during the intermediate precision study, the standard deviation for intermediate precision, when combined with the uncertainty due to bias, can be used as the initial standard uncertainty for the method.

Discussion

The estimation of measurement uncertainty is discussed in the Eurachem/CITAC Guide *Quantifying Uncertainty in Analytical Measurement*[v]. Any factors that could impact the uncertainty must be accounted for and included in the uncertainty estimate. Such factors are the matrix of the sample, the difference between the certified reference material and the sample, or sample preparation, etc.

The uncertainty estimated from the method validation can be compared to the target uncertainty to determine if the method is fit for use. The uncertainty derived from the method validation work is only a preliminary estimate. The initial estimate will have to be revisited as the laboratory gains more experience with the routine use of the method on real samples.

Conclusion

Method validation is a key activity to producing results that are fit for purpose. No matter the purpose, the data has to have adequate accuracy (bias and precision) so that decisions based on the results are not made in response to random analytical variation. Target measurement uncertainty allows unambiguous, quantitative expression of the bias and precision requirements for the method and for the validation of that method. The target measurement uncertainty can be derived from the decision rule based on the use of the data.

The analytical parameters for the method validation must be understood, clearly identified and adequately demonstrated in the validation of the method. The experiments needed to demonstrate method performance are now well understood and readily designed. The instruments must be qualified so that their performance is known and controlled.

In today's analytical world, there is now a clear process to define the purpose of a method through the decision rule, to calculate the target measurement uncertainty, and to derive the analytical parameters and their acceptance criteria, in order to validate a method. This will ensure data that is fit for purpose.

References

[i] V J Barwick and E Prichard (Eds), *Eurachem Guide: Terminology in Analytical Measurement – Introduction to VIM 3* (2011). ISBN 978-0-948926-29-7. Available from http://www.eurachem.org/index.php/publications/guides

[ii] ISO/IEC 17025:2005, General requirements for the competence of testing and calibration laboratories, International Organization for Standardization (ISO), Geneva. http://www.iso.org/iso/home.htm.

[iii] *H Ramsey and S L R Ellison (eds.) Eurachem/EUROLAB/CITAC/Nordtest/AMC Guide: Measurement uncertainty arising from sampling: a guide to methods and approaches Eurachem (2007)*. ISBN978 0 948926 26 6. Available from the Eurachem secretariat http://www.eurachem.org/index.php/publications/guides

[iv] *International vocabulary of metrology – Basic and general concepts and associated terms (VIM)*, JCGM 200:2012, Joint Committee for Guides in metrology (JCGM), 2012, www.bipm.org.

[v] S L R Ellison and A Williams (Eds). Eurachem/CITAC guide: *Quantifying Uncertainty in Analytical Measurement*, Third edition, (2012) ISBN 978-0-948926-30-3. Available from http://www.eurachem.org/index.php/publications/guides

[vi] V J Barwick, S L R Ellison (Eds), *VAM Project 3.2.1 Development and Harmonisation of Measurement Uncertainty Priniciples Part (d): Protocol for uncertainty evaluation from validation data*. (January 2000) Version 5.1 LGC/VAM/1998/088

[vii] M.L. Jane Weitzel, Wesley M. Johnson, Accreditation and Quality Assurance: Journal for Quality, Comparability and Reliability in Chemical Measurement, *Using Target Uncertainty to Determine Fitness for Purpose*, Volume 17, Issue 5 (2012), Page 491-495, (DOI)

10.1007/s00769-012-0899-x

[viii] ASME B89.7.3.1-2001 *Guidelines for Decision rules; Considering measurement Uncertainty in Determining conformance to Specifications*; copyright 2002 by the American Society of Mechanical Engineers

[ix] John Keenan Taylor, *Quality Assurance of Chemical Measurements*, Lewis Publishers, 1987, p. 266

[x] James N. Miller, Jane C. Miller, *Statistics and Chemometrics for Analytical Chemistry*, 6th Edition, 2010, Pearson Education Limited

[xi] Joseph M. Juran, A. Blantom Godfrey (Eds), *Juran's Quality Handbook* Fifth Edition, McGraw-Hill – International Edition, 2000.

[xii] United States Pharmacopeia General Chapters: <1058> *Analytical Instrument Qualification*, http://www.uspnf.com/uspnf/pub/ (accessed March 9, 2012).

[xiii] GAMP® *Good Practice Guide: Validation of Laboratory Computerized Systems*, April 2005, http://www.ispe.org/gamp-good-practice-guide/validation-laboratory-computerized-systems (accessed March 9 2012)

[xiv] AOAC International, *How to Meet ISO 17025 Requirements for Method Verification*, http://www.aoac.org/accreditation/accreditation.ht (accessed March 9 2012).

[xv] United States Pharmacopeia General Chapters:<1225> *Validation of Compendial Procedures*, http://www.uspnf.com/uspnf/pub/ (accessed March 9, 2012).

[xvi] Eurachem Guide: *The fitness for purpose of analytical method: A laboratory guide to method validation and related topics*, 1998, ISBN 0 948926 12 0, www.eurachem.org (accessed March 9 2012)

[xvii] *Reference materials – contents of certificates and labels*, ISO Guide 31:2000, International Organization for Standardization (ISO), Geneva.
Calibration in analytical chemistry and use of certified reference materials, ISO Guide 32:1997, International Organization for Standardization (ISO), Geneva.
Uses of certified reference materials, ISO Guide 33:2000, International Organization for Standardization (ISO), Geneva.
General requirements for the competence of reference material producers, ISO Guide 34:2009, International Organization for Standardization (ISO), Geneva.
Reference materials – General and statistical principles for certification, ISO Guide 35:2006, International Organization for Standardization (ISO), Geneva.
http://www.iso.org/iso/home.htm.

[xviii] Wernimont, Grant T., Evaluating the Ruggedness of an Analytical Process. *Use of Statistics to Develop and Evaluate Analytical Methods*; AOAC International: United States of America, 1985; pp 78-82

[xix] International Union of Pure and Applied Chemistry, Analytical Chemistry Division, Commission on General Aspects of Analytical Chemistry, *Selectivity in Analytical Chemistry* (IUPAC Recommendations 2001. Pure Appl. Chem. 2001, 73, 1381-1386

[xx] USP Medicines Compendium, http://www.usp-mc.org/

[xxi] NMKL Procedure No. 20(2007) *Evaluation of Results from Qualitative Methods*, Nordic Committee on Food Analysis, Available from: http://shop.nmkl.org/

[xxii] Corula Wilrich; Peter-Theodor Wilrich *Estimation of the POD Function and the LOD of a Qualitative Microbiological Measurement Method*, Journal Of AOAC International, Vol. 92, No. 6, November 2009, 1763-1779

[xxiii] AMC Technical Brief No.13, September 2003, *Terminology – the key to understanding analytical science. Part 1: Accuracy, precision and uncertainty,* Royal Society of Chemistry, 2003
http://www.rsc.org/images/brief13_tcm18-25955.pdf

[xxiv] *Accuracy (trueness and precision) of measurement methods and results -- Part 2: Basic method for the determination of repeatability and reproducibility of a standard measurement method,* ISO 5725-2:1994 International Organization for Standardization (ISO), Geneva.

[xxv] S. L. R. Ellison, Vicki Barwick, Trevor J. Farrant *Practical Statistics for the analytical scientist: a bench guide,* Royal Society of Chemistry, 2009

[xxvi] Larry B. Barrentine *An Introduction to Design of Experiments A Simplified Approach,* ASQ Quality Press, Milwaukee, Wisconsin, 1999.

[xxvii] ML Jane Weitzel (2011) *The estimation and use of measurement uncertainty for a drug substance test procedure validated according to USP <1225>* Accreditation and Quality Assurance: Journal for Quality, Comparability and Reliability in Chemical Measurement, Volume 17, Number 2 (2012), 139-146, DOI: 10.1007/s00769-011-0835-5

CPSIA information can be obtained at www.ICGtesting.com
Printed in the USA
LVOW08s0354260713

344762LV00001B/192/P